God and Quantum Physics: The Philosophy

Table of Contents

Introduction

The idea of bringing the weirdest aspect of modern science, i.e. Quantum Mechanics, which was a great nightmare for one of the greatest minds in human history; Religion, which is unarguably one of the major culture of every human society; and Philosophy, meant for all but accepted by those with open minds and rejected by the rigid ones, to a point of contingence is the craziest idea anyone could ever think of. This seems not to be plausible but it is interesting to understand that everything about life is connected, it will just require another great mind to find out how.

For better understanding, the connectivity of everything about life can be simulated as a mathematical Venn diagram of a universal set consisting of different sets of variables in which every set is either a subset of another set or intersects with another set. In other words, no set or variable is existing independently in the universal set... Of course, placing this analogy into our imagination will reveal a kind of rough unorganized Venn diagram which can be confusing but that is it, the mystery of life, everything is connected though strange and confusing.

In a more practical sense, the connection of everything can easily be observed within the human environment which is everywhere to see. A thought about the respiratory interaction between animals and plants can be used as an example. Animals take in oxygen from the atmosphere, make use of it, and produce carbon dioxide which is expired. The carbon dioxide is taken up by the plants, use it for photosynthesis in the presence of UV light in the chloroplast to produce food, for animals, and oxygen which is taken up by animals again for respiration. In actuality the chloroplasts are descendants of a single-celled organism which existed since about 3 billion years ago called cyanobacteria which, for the sake of survival during the course of evolution, established a symbiotic relationship with the plant cell. At that time, the earth is composed of the same amount of nitrogen that we have now but almost no oxygen with larger amount of carbon – dioxide, making the earth inhabitable. So, the cyanobacteria gradually and slowly coverts the carbon dioxide molecules into oxygen via photosynthesis using the energy from the sun for about 2.5 billion years and just about 900 million years ago, oxygen starts to build up in the atmosphere before the ozone layer could be formed in about 600 million years ago. The formation of the ozone layer makes it possible for earth to support complex multicellular organism like plants and animals. It is quite interesting that the chain of survival of animals can be traced back to these events that occurred few billion years ago.

However, there are still numerous phenomena which shows that everything about life is connected. Another example is the story of the element, Iron (Fe). In the beginning, the universe contains no iron except Hydrogen and Helium. However, Iron was only able to form through the explosion and collision of stars during the course of galactic collisions. Mind you, this element is the central heart (constituent) of the hemoglobin molecule in the human blood which is very important for the blood's ability to bind with oxygen. Imagine, the Iron atom in human blood is connected to what happens in space? It is ridiculous but in whatever way it is, there is a connection there.

Likewise, consider the water cycle, the sun takes up water from the earthly water bodies and plants by evaporation to the sky, the water molecules condenses, form clouds, then rain falls which is taken up by plants from the soil, make use of it for proper growth which in turn serves as food for animals and humans, then the cycle goes again. So with this, we are able to understand that, one way or the other in life, one thing leads to another, gradually making everything look like a junk coils of a rope without a beginning or an end. It is strange but true.

In spite of these, connecting all that we understand about life in different perspectives deserves a shot. It is all about asking the right questions and finding the appropriate answers to them. First, Science has enormously succeeded in unraveling most of the mysteries about life leading to theories and models to explain how everything works so much that it is arguably claimed that it provides the answers to all questions any human could raise. The understanding of how the universe came into being, knowledge of the galaxies, planets, stars, other celestial bodies and how they interact, the study of all the so-called living things, micro-organisms and even human himself, animals, plants, the knowledge of atoms, molecules and how they form the building blocks of everything, just to mention a few, are now known and all thanks to science. With these knowledge, humans have been able to understand and strive to survive in his environment. The broad scope of science requires specialization leading to its different branches, mathematics, physics, chemistry, biology, medical sciences and so on.

Contrary to what some might think, it is not incorrect to posit that physics is the king of science as it studies the basic fundamental nature of everything and it uses the language of mathematics, thus the queen of science.

The superiority of physics over other sciences is due to the fact that it is a study towards understanding the behavior of everything. Everything that exists is matter, as far as it takes up space and weighs. Physics studies the fundamental nature of matter, thus giving us an understanding of everything. However, physics classifies matter into two broad scales, which we can regard as the 'large' and 'small' scale and the behavior of matter in each scale is guided by completely different laws. The laws of classical physics explain the behavior of matter at the large scale which is easily observable around our environment. For example, the laws that guide the behavior of a rolling ball, a falling apple, a moving car, stars, planets and so on. Meanwhile, at the 'small' scale i.e. at the subatomic microscopic level, the behavior of matter is guided by the laws of quantum physics, like the understanding of the motion of electrons in an atom.

Quantum physics is undoubtedly annoying and absolutely absurd to the intuitive sense but nevertheless, it has caused a great paradigm shift in the way we perceive the universe and it poses a potential to answer some of the mysterious questions about the universe.

Second, every human society is cultured with a particular way of life and practice which they are devoted to, called Religion. Religion molds the human ideology and perception about life as it shapes his beliefs and convictions which are derived from his understanding about everything as explained by his religion. In other words, religion also gives an explanation behind everything like its origin, reason behind its existence and also propose its fate, thus making it a very important aspect of life. However, the theory on the origin and fate of everything and how everything works as proposed by religions is centrally based on the concept of a Supreme Being, God. This concept proposes the existence of a Supreme (supernatural) Being, God, who created life and possess sovereignty over everything. There are different kinds of religion due to several beliefs and convictions about God but we have two major groups of human based on the acceptance of this concept; Theists and Atheists. Atheists reject the idea of the existence of any Supreme Being whatsoever and believe that everything must have existed by chance whilst Theists accept the idea of the existence of a Supreme Being but due to the inadequate understanding of 'who', 'what', 'where' and 'why' God(s) is/are results into different beliefs among theists. Nevertheless, the concept of a Supreme Being with sovereignty over everything determines the way of life of humans, hence a very important paradigmatic view of humans about life.

Third, the inherent nature of human to search for the understanding of life leads him to another crucial perspective of understanding everything, dubbed as Philosophy. Principally, with the intellectualistic nature of human, philosophy poses questions about everything, provide answers out of reason, in order to understand everything about life. To achieve this, philosophy questions the explanations behind everything as provided by the scientific and religious paradigmatic view of life in order to make the appropriate sense out of them. In other words, philosophy is the rule that guides the understanding of everything in whatever the human perspective is, including itself.

It is quite crazy when one questions himself but on the second thought, it is safer to be double sure than to realize that one is a fool at the end.

However, since everything about life is connected then perhaps, all that we understand about life could also share a connection. In other words, the different paradigmatic views of life that tells us all that we understand about everything, science, religion and philosophy could share the same space. After all, they all explain the same thing, life. As stated in the first paragraph, this seems more implausible and quite crazy but come on, there was a time no one knew the earth was oblate and not flat.

What is Life?

Consider a man with a pure clean slate mind, just like a baby given birth to, a second ago without the slightest idea of anything. Such mind doesn't know what is 'what' nor understand what 'what' is. Nevertheless, the mind is characterized with curiosity and thirst to know what is 'what' and understand what 'what' is, including itself. Therefore, using this analogy, we take backward steps to when our minds was a pure clean slate without any slight knowledge of anything. This approach is adopted to explain how we came to understand all that we have understood about life, posing the question, how do we know what everything is and how it works? Thus, explaining how everything adds up to bring about existence called life. In spite of this, man seeks to understand everything around him (including himself) out of curiosity, by providing answers to the questions of what everything is, why everything is and how everything is, to reveal the secret behind existence.

This search for knowledge of everything was termed as 'Philosophy' by an ancient Greek thinker, Pythagoras in the 6th century BCE. Since then, philosophy has grown over the years into different eras of ancient, medieval and modern philosophical stages across every part of the world, to pose logical questions about everything and provide answers to them through, logical reasoning. Historically, philosophy is associated with all forms of knowledge but right from the ancient ages till the 19th century, philosophy was divided into natural, moral and metaphysical philosophy.

Natural philosophy involves the study of nature of the physical world around man. Moral philosophy entails the study of what is good, right, wrong and virtues. Metaphysical philosophy deals with the study of existence and the being of everything.

Meanwhile, philosophy has been extensively branched in modern philosophy with natural philosophy yielding natural sciences like physics, biology, chemistry, cosmology, astrology and so on to understand the nature of everything i.e. to know what 'what' is, made up of and how it behaves.

Moral philosophy brought about ethics that deals with the understanding of values i.e. what is right and wrong or good or bad. Also, aesthetics, that involves the study of art, beauty and music and political philosophy that entails the understanding of how to attain possibly, a utopian state.

Metaphysical philosophy splits into some formal sciences like mathematics, the study of reasoning referred to as logic;

epistemology that tells the understanding of how we know what we know; ontology, which deals with the study of being i.e. what makes a being out of being or rather the essence behind being or existence.

All these encompass the study of understanding everything about life telling us all what, 'what' is, why 'what' is, how 'what' behave, if 'what' is right, wrong, good or bad, how 'what' even exist, how to interact with 'what' and so on, therefore giving us the idea of all that we know about life and even live it.

Without any doubt, philosophy is a very important paradigmatic view that determines the way we understand all that we understand about life by encouraging the search for the answers to all the questions about life through rational arguments and logical reasoning.

In the ancient Greek philosophical era, a great thinker, Aristotle encouraged the idea that the observation of physical phenomenon could lead to the discovery of the natural laws that governs them. In the 4th century, he wrote his first work in respect to that line of study later known as Science. Science stems from natural philosophy of the ancient but exist quite independently of the modern day philosophy. The main scope of science is to study the fundamental nature of the observable physical world and understand the natural laws that guides their behavior. However, with the goal to understand the nature of everything about life, science also includes several branches of specialization like physics, mathematics, biology, astronomy, cosmology, psychology and so on. Nevertheless, science aims to understand the basic constituents of everything. Which are simply matter and energy, so every branch of science is restricted to the study of the nature of a particular aspect of everything about the physical world except physics which does not only study the fundamental physical nature of everything but also describes the motion, interaction and laws that guide the basic constituents of everything i.e. matter and energy and therefore referred to as the 'King of Science'.

In fact, it is historically claimed that other branches of sciences emanated from physics with respect to their specializations to exist independently and therefore limited to their respective domains. Moreover, physics has successfully been able to describe the natural laws that guides physical phenomena through the language of mathematics such as geometry, calculus algebra and so on. Hence, it is safe to refer to mathematics as the physics' romantic partner, making it the 'Queen of Science'.

Every physical thing takes up space and weight, giving us the definition of matter and its behavior or motion caused by an agent called energy. In Aristotelian physics, he proposed that all matter is composed of five main elements; "Earth", "Water", "Air" and "Fire" due to the four sensible features; hot, cold, wet and dry as proposed earlier by another ancient Greek philosopher, Empedocles and added "Aether", as the divine substance of heavenly spheres of stars and planets. This was after a time when another great thinker before Aristotle in the pre-Socratic era, Thales, inferred that everything is made up of "water". Thereafter, an early physicist, about half of the 5th century, Leucippus, strongly disagreed with the idea of divine intervention and claimed that every natural phenomenon is directed by a natural cause. Then his student, Democritus came about the notion that everything is completely composed of several imperishable, indivisible elements called atoms which is dubbed as the 'theory of atomism'. Nevertheless, the study of the nature of matter and search for the understanding of its behavior, specifically its motion continues to progress with the likes of Archimedes, Aristarchus of Samos, Hipparchus, Ptolemy and so on till the days of Galileo Galilei and Sir Isaac Newton. Galileo was regarded as 'the father of modern physics' and even 'the father of modern science' due to his major contributions to physics, astronomy to be precise and he believed that the motion of matter, whether naturally or artificially, can't just only be derived and be described mathematically but universally constant. He was known for his invention of hydrostatic balance and work on the center of gravity of solid bodies. Stephen Hawking once said, "Galileo, perhaps more than any single person, was responsible for the birth of modern physics"

During the 17th and 18th centuries, Isaac Newton made great achievement in mechanics and astronomy by discovering a single system to understand how everything works. In his *Philosophiae Naturalis Principia Mathematica* (Mathematical Principles of Natural Philosophy), published in 1687, he presented his three laws of motions that explains the relationship between matter and motion due to energy and also the law of universal gravitation which explains the behavior of a falling matter and that of planets and other celestial bodies using a completely novel mathematics, arguably invented by him called *Calculus*. This helps us to understand and predict how matters could behave through mathematical laws as presented by Newtonian physics. He also contributed greatly to Optics, the study of light and all of these are regarded as the classical physics as they explain the behavior of matter at the large scale of dimension. Also, this invariably marks the beginning of modern physics.

However, the idea that matter consists of indivisible and indestructible elements called atoms was upheld till the 18th century which saw John Dalton to improvise on the works of Leucippus and Democritus to propose his atomic theory until the 19th century. Although, Henri Becquerel discovery of radioactivity in 1891 suggested that atoms are not indivisible and indestructible but can disintegrate forming different atoms and William Crookes' discovery of the cathode rays in electric discharge tubes revealed that atoms contain negatively charged particles. Sir J.J. Thompson named the negatively charged particle 'electron' in 1897 as a component of atom and due the fact that atoms are neutrally charged, it was logical that atom couldn't just consist of only electrons, leading Ernest Rutherford to the discovery of 'proton', a positively charged particle in the nucleus situated at the center of an atom with the electron revolving round it. This is referred to as the planetary model of atom. Later on, another elementary particle with no charge called 'neutron' was discovered which is also located in the nucleus. Interestingly, recent discoveries have shown that the nucleus is not just a combination of protons and neutrons, but can be broken down into more elementary particles like quarks which comes in different flavors and colors – three quarks make up a proton and a neutron each, gluons, muon, muon neutrino, Higgs boson, W boson, Z boson, photon and so on. In fact, to our wildest imagination, it was a great amazement when the equation of another great English mind, Paul Dirac, which suggests that every matter has a twin partner called anti-matter. An anti-matter is just the same like a matter but only with an opposite charge i.e. the anti-matter of a positive charged matter will bear a negative charge instead but will be just the same with the matter in other terms like mass. For instance, every particle has its own anti-particle as, a "positron" is an anti-particle of an electron, same as anti-proton for a proton. Even though, there is an imbalance in the amount of matter and anti-matter in the universe such that there is a lesser amount of anti-matter than matter, a collision between a matter and its anti-matter like an electron and positron, will result into annihilation i.e. the release of nuclear energy. So perhaps, for every 'you' there is an 'anti-you' somewhere, in which if you collide together, you will annihilate and release a great amount of energy that could just be enough to destroy the whole Earth!

A Danish scientist, Neils Bohr, in 1913 discovered that electrons move around the nucleus in a specific circular orbit called shells and the energy possessed by an electron in an atom does not vary but restricted and limited to a number of discrete or packet of values and therefore quantized. Prior before that, Max Planck had also observed that the radiations emitted from substances involves the emission of energy in discrete packets which is referred to as *Energy Quanta* in 1902. Building on the idea of Max Planck, Albert Einstein explained the concept of photoelectric effect in 1905 and proposed that light waves are in tiny bundles and packets of energy known as photons or quanta which causes the emission of electrons when the waves hit a material. However, the understanding of the photoelectric effect reveals that matter can behave both like a particle and wave. By this, we mean that matter can exist like a physical entity i.e. like an object or spreads out in form of energy as waves, just the like the ripples of water observed when an object is dropped in a water body. This is called the 'wave – particle duality' nature of matter. With this, we are able to understand what is happening in the atomic world as the electrons are observed to be in a very rapid motion around the nucleus such that it is very difficult to measure the exact position of the electron, leading another physicist Erwin Schrödinger to bring about the concept of *orbital*, using the mathematical elegance of calculus and complex numbers, taking account of the quantum states and wave-particle nature of matter. Orbitals indicates the regions in an atom around the nucleus where the probability of finding an electron is the greatest. In spite of these, the weirdest aspect of modern science was born, 'Quantum Physics' which explains the nature of everything at the fundamental subatomic level.

Neils Bohr is regarded as the 'Father of Quantum Physics' and since then, the study of quantum physics has grown rapidly with several achievement from many physicists unraveling the most mysterious secrets of life, therefore causing a significant paradigmatic shift in our understanding about everything.

Life is not just about the physical nature of everything alone but also includes the supposedly abstract aspect that involves how to live life, itself. That is, how to interact with the physical world around us, to understand the rules or laws that guides the interaction of everything in the universe. This brings about our way of life, the concept of how we conscientiously interact with the physical world around us and among ourselves called Religion.

Religion is a fundamental culture of every human society. In fact, it is safe to claim that it has been in existence since the day the first human had set his foot on the Earth because it entails all forms of practice that human is seriously devoted to, through the course of his existence.

In spite of this, it is crucial to understand how we came about the rules or laws that guides this aspect of life as it is another major paradigm through which we perceive life due to the fact that it molds our ideology, shaping our beliefs and convictions about everything.

Every society is cultured with its own way of life, leading to the existence of different beliefs and convictions about life, hence the reason for several religions. But generally, a religion is characterized with rules and laws that guides how a group of people live their life and how they interact with the physical world around them, reflecting their understanding about it. These rules and laws are usually claimed to be divine and attributed to a supernatural being, God, who is understood to be the creator, the wisdom behind, and possessor of sovereignty over everything.

This concept of God is central to every religion around which all other ideas are built around. In this paradigmatic view of life, God is seen as the Omniscient that understands everything about life and therefore, stipulates the appropriate rules and laws that guide interaction between everything which He sent to humans through His chosen ones among them called Prophets. Thus, forming a major perspective of how we knew all that we understood about everything.

In brief, all our understanding of everything about life can be categorized into three major paradigmatic perspectives: Philosophical, Scientific and Religious perspectives. Each, in its own way explains to us what life is.

Connecting the Dots

Apparently, the ultimate goal is to proffer a unique and concise paradigm which will encapsulate the understanding of everything about life.

Although, it is understood that this is the individual aim of the three major perspectives through which we understand life presently. But at some point, they disagree such that they lose the connection in the course of their search for the understanding behind everything. As a result, different assertions were made leading to diverse opinions upheld among humans about the understanding of life.

The philosophical perspective gives the freedom to pose questions about everything and provide answers through rational arguments and logical reasoning since human is characterized with the power of intellect, giving room for relativity in the proposition of verdicts, leading to diverse opinions. But this is only embraced by the open minded and totally rejected by humans with a fixed mindset based on what they have grown to believe plus the claim and belief that the human intellect is restricted and unable to conceive or understand some things. This ideology of a rigid mind, is mostly influenced by the religious perspective of understanding life which limits the freedom granted by philosophy, banking all its ideas and understanding on divine rules and laws which must not be violated as set by the omniscient supernatural being, God. Hence, creating a major loop hole there between the two perspectives.

Moreover, despite the advancement and great achievement of science over the years, creating the easy technological world around us, some humans find it difficult to embrace its theories and laws that give the understanding of the physical universe. It is surprising and quite absurd that humans embraces the applications of these theories, like computers and mobile phones but rejects the theories behind them. The cause of this effect can also be exerted on the socio-cultural system of religion which criticizes the flexibility of science. Flexibility in the sense that, just like philosophy, science holds the claim that the understanding of everything can be explained by natural laws and without any divine intervention. In other words, science is like that stubborn little child that believes that the fact that he can't explain a phenomena today does not mean he will never be able to. It is a stubborn belief but yet, it is unarguably the fuel behind the progress of science over the years. In fact, who could have ever thought that there are billions other planets in our universe apart from Earth, understand that the universe is not static but expanding and galaxies are moving away from us, that we are made up of particles as tiny as electrons and understand how crucial they are to our existence or understand that at a time in the past everything was jammed together in a very tiny, hot and point in space?

Philosophy also questions scientific methods and strives to interpret scientific results to attach the appropriate meaning to them. Even though science was a major branch of philosophy as natural philosophy in the ancient Greek philosophy, it was eventually extracted from philosophy due to the fact that they employ different methods in their quest for the understanding of everything about life. Science firmly and solely agrees with experimental and empirical data. If it does not agree with the experiment or cannot be tested, then according to science, it is bollocks. Whereas, philosophy still seems to be more flexible with her methods and holds that, even if it does not agree with experiment or cannot be tested, it does not necessarily mean that it does not make sense and in fact, can even go as far as questioning the reliability and accurate precision ability of the experimental process. Therefore, creating another major loophole between them.

Nevertheless, it is necessary that these perspectives are brought together into terms in order to give a general paradigm of understanding of everything about life. This sounds crazy but in fact, it is the only way we can affirmatively assert that we have succeeded in understand everything. The plausibility of this idea is questionable because each perspective has progressed independently over the years. However, they are all driven towards the same aim; understanding life. If truly, the philosophical and scientific methods of understanding life plus the concept of religion are right, then it is rational to deduce that they can all share the same space or arrive at a contingent point. The inability to bring these perspectives together is a logical proof that we have gone wrong somewhere and thus failed in our understanding of life.

In spite of this, the idea of taking each perspective in turns and connecting the dots deserves a shot. With that, we will be able to understand their differences and perhaps solve the discrepancies, bridging the gap between them to reach a point of contingence.

Religion and Philosophy

The first major hint of comparison between religion and philosophy came from an Islamic philosopher in the medieval ages of philosophy, Al-Farabi, who adopted the philosophy of Aristotle and Plato and asserted that religion is an imitation of philosophy because it is limited to the image of essence. Essence, according to the ancient Greeks, is the quality or set of qualities that make a thing what it fundamentally is, and which it must necessarily have to be what it is or it loses its identity. To Al-Farabi, religion is a depictive imagination that can only be like the essential reality i.e. the real image of essence known by philosophers but can never be like it. In his book, Al-Farabi's Philosophy of Plato and Aristotle, part I: The Attainment of Happiness, he wrote:

> Now when one acquires knowledge of the beings or receives instruction in them, if he perceives their ideas

themselves with his intellect, and his assent to them is by means of certain demonstration, the science that comprises these cognitions is philosophy. But if they are known by imagining them through similitudes that imitate them, and assent to what is imagined of them is caused by persuasive methods, then the ancients call what comprises these cognitions, religion. And if those intelligibles themselves are adopted, and persuasive methods are used, then the religion comprising them is called popular, generally accepted, and external philosophy. Therefore, according to the ancients, religion is an imitation of philosophy.[1]

More explicitly, Al-Farabi noted that philosophy does not just entail the grasp of the real nature of essence itself, it also describes it whilst religion fails in this regard. In an introduction to the book by Muhsin Mahdi, who translated the book from the Arabic text, a definition is ascribed to religion as:

"…an imitation of philosophy in the restricted sense in as much as both comprises of the same subjects and both give account of the ultimate principle of the beings or in so far as religion supplies an imagination account of, and employs persuasion about, things of which philosophy possess direct and demonstrative knowledge."[2]

However, Al-Farabi bluntly placed philosophy above religion as he believed that a philosopher has a more grasp of understanding on the essence of everything than other people who in fact, do not have the ability to understand the essence of everything as the philosophers. Although, another Islamic philosopher, Al-Ghazzali strongly disagreed with this in his book, The Incoherence of the philosophers. Nevertheless, Al-Farabi was able to point out the similarities and major difference between religion and philosophy. Both are similar in the sense that they both account for the essence of everything and differ due to the fact that religion is restricted certainly by imaginative depiction of reality as opposed to philosophy.

Taking it literally for the benefit of understanding, religion and philosophy account for the essence i.e. understanding of everything in a concept of the same structure. Religion is a system with a concept that involves the understanding of being and existence of everything, a similitude to the idea of "ontology" in philosophy; understanding of the rules and laws that guides the interaction of everything which is equal to "ethics" in philosophy; derived from the message sent by a supernatural being through His chosen ones among humans documented in scriptures, relating to study of how we know in philosophy called "epistemology".

Religious concept also entails the idea of existence of paranormal beings like God and angels which under philosophy can be liken to "metaphysics" and involves an ideology and way of reasoning based on historical and sacred events and stories in the scriptures, thus seems like "logic", the study of reasoning in philosophy.

Therefore, religion and philosophy enjoys the same thing in common more than what we can realize on the surface. But it is worthy of observation that the religious concept is confined under a given fixed sets of rules and laws that explains and guides the understanding of everything documented in a sacred scripture stated by a supreme being , God, who created and has sovereignty over everything, thereby restricting the freedom given to human's intellect by philosophy. Philosophical concept is characterized with limitless freedom of wit to bring about rational arguments in order to discover the best answers to questions about everything to understand them. In spite of this, philosophical questions have been raised by philosophers on the concept of God itself, which is obviously and unarguably central to the religious perspective. The philosophical argument of the existence of God is regarded as the ontological argument. At this point, the relationship between philosophy and religion becomes unhealthy and collapses.

An idea of the ontological argument surfaced in the works of Xenophanes, an ancient Greek philosopher and some works of Parmenides, Plato and Neo-Platonists also show an argument of such. Plato even asserts in one of his works, Timaeus that, "the cosmos was created and not eternal even though, it was framed by the creator after an eternal, unchanging model". Aristotle also made his contribution to the concept of the existence of God in his philosophy of theology where he argued that there are two major attributes of being: actuality and potentiality.

Actuality is characterized with perfection, realization and fullness of being i.e. determining the determinable principle while potentiality involves imperfection, incompleteness i.e. the determinable principle or that to be determined. According to Aristotle, all material things possess some potentiality. He also coined the term, 'unmoved movers' who possess actuality, "actus purus" and hence they are unchanging, eternal, immaterial substances. Therefore, he asserts and argues that:

> "There can be no pure potentiality without any actuality whatsoever. All material substances have unactualized potentials... Although motion is eternal, there cannot be an infinite series of movers and of things moved. Therefore, there must be some, who are not the first in such series, that inspire the eternal motion without themselves being moved as the soul is moved by beauty"[3]

Avicenna (Ibn Sina), a philosopher of the medieval ages, who built on the works of Aristotle and Plato, also presented his own argument which, just like the claims of Aristotle, distinguishes between a thing that need an external agent or cause to exist which he termed as 'a contingent thing' and a thing that must exist due to its essence or inherent nature referred to as 'a necessary thing'. He argued that everything in life is a contingent thing that needs external agent to cause its existence which also will need another cause to exist and so on till it leads to an infinite regress. For example, a wooden table is made by a carpenter who caused the table to exist by carving the wood, the wood was caused to exist from trees, and also the tree is also caused to exist from who planted it, who is also caused from another being and so on. The same also goes for the carpenter himself who is also caused to exist from his parent who are also caused to exist from his grandparents and it goes on till infinity. But instead to reach an infinity regress, that is to go on and on with the chain of existence, Avicenna concluded that at a point, some necessary cause, such as God is needed to end the chain. Mulla Sadra, an Iranian Shia Islamic philosopher whose philosophy was influenced by the works of Ibn Sina (Avicenna), Ibn 'Arabi and other Muslim philosophers posed his argument on existence of God in Al-asfar al-arba'a (four journeys) wherein he wrote that:

> Existence is a single, objective and simple reality, and there is no difference between its parts, unless in terms of perfection and imperfection, strength and weakness… And the culmination of its perfection, where there is nothing more perfect, is its independence from any other thing. Nothing more perfect should be conceivable, as every imperfect thing belongs to another thing and needs this

other to become perfect. And, as it has already been explicated, perfection is prior to imperfection, actuality to potency and existence to non-existence. Also, it has been explained that the perfection of a thing is the thing itself, and not a thing in addition to it. Thus, either existence is independent of others or in need of others. The former is necessary, which is pure existence. Nothing is more perfect than Him. And in Him there is no room for non-existence or imperfection. The latter is other than Him, and this is regarded as His act and effect, and for other than His there is no subsistence, unless through Him. For there is no imperfection in the reality of existence, and imperfection is added to existence only because of the quality of being caused, as it is impossible for an effect to be identical with its cause in terms of existence.[4]

However, the ontological argument was primarily developed by Anslem of Canterbury in his work, *Proslogion* in 1078, wherein he defined God as a "being than which no greater can be conceived" and posit that since the concept of God can be conceived, even to a 'fool', then it exist in the mind and therefore exist in reality. He argued that since it is true by definition that God is a being which no being greater than can be imagine, therefore existing in the mind. And that a being that exist in the mind and also in reality is greater than a being that exists only in the mind, therefore if God exists only in the mind, then we should able to imagine something greater than God i.e. there is a possible existence of a greater being. But we cannot imagine something greater than God in the mind, hence he must exist in reality.

French philosopher, Rene Descartes, also follow suit and wrote in the Fifth Meditation:

But, if the mere fact that I can produce from my thought the idea of something entails that, everything entails that, everything that I clearly and distinctly perceive to belong to that thing really does belong to it, is not this a possible basis for another argument to prove the existence of God? Certainly, the idea of God, or a supremely perfect being, is one that I find within me just as surely as the idea of any shape or number. And my understanding that it belongs to his nature that he always exists is no less clear and distinct than is the case when I prove of any shape or number that some property belongs to its nature.[5]

Gottfried Wilhelm Leibniz partially goes with Descartes by questioning the notion of absolute perfection he attributed to the Supreme Being. According to Descartes, all forms of perfection coheres or unifies into the Supreme Being but Leibniz argued that the coherence of a supreme perfect being is difficult and impossible to analyze or demonstrate even though he still felt that Descartes' argument is valid. All of these do not leave out the modern philosopher, Immanuel Kant, who upheld the belief that in so far as a thing has the possibility to exist, then there must be a rationale behind the possibility of its existence and that, he attributed to a single necessity identified as being God by him. Kant critically tries to demonstrate that this being is characterized with omnipotence, omniscience and omnipresence.

The critical approach of Kant and Leibniz's struggle to demonstrate the coherency of the supreme perfect being, points the weakness in the argument presented by the earlier philosophers. Although, the earliest criticism came from Gaunilo of Marmoutiers who had some objections to Anslem's argument, so also is Thomas Aquinas who claimed that since human does not know the nature of God and therefore cannot be imagined or conceived as expressed by Anslem. The inability to unify the omnipotent, omniscient and omnipresent characteristics of God makes the idea of a supremely perfect being incoherent. Putting it simple, if God is omnipotent, he should be able to create a being with free will and if at the same time, He is omniscient, He will possess the knowledge of all and thus He will be able to know with certainty what such being will do, technically dispossessing the being of his free will. This explains the problem of incoherence of a supreme perfect being as the qualities required for the greatest being cannot coexist in the one being at the same time and therefore cannot exist.

As a result, philosophy produces two broad classes of beliefs on the existence of God: Theism and Atheism. Theism accepts the belief in the existence of God while the rejection of the belief is Atheism. However, Al-Farabi's explanation of religion in *Al-Farabi's Philosophy of Aristotle and Plato*, helps us to understand that we can have different kinds of religion. Since religion only involves the imaginative description of essence of everything and not the essential reality, then we can have different images of essence and thus different religions.

Although, he claimed that some images, that is some religions are better than others. Not exactly the same, but in view of this, there are different forms of theism as a result of diverse beliefs and conviction about God who, as whole in the context of theism, is believed to exist. There are numerous kinds of theism, though broadly we have 'monotheism' and 'polytheism'. The former is the belief that only one God exist which include modern beliefs like Christianity, Islam, Judaism, Sikhism, and Zoroastrianism and the latter is the belief that there is more than one god such as Henotheism, Kathenotheism and Monolatrism.

A view of theism also looks into connection of God and physical world. Pantheism is the belief in the equivalence of physical world to god and on the other hand, there is Panentheism, the belief that the physical world is intertwined to god or gods. Another understanding of God is that He created the world but afterwards, does not have any effect in its course of existence and therefore devoid of the belief in miracles, prophecies or any form of divine intervention, this is referred to as Deism. From this, we have 'Pandeism' that suggest that God created everything and afterwards becomes equal to it and 'Polydeism' that posit that there are many gods but do not interact with the universe. Nevertheless, there is also a belief that the existence of God is unknown and cannot be known. This is called Agnosticism.

In spite of all, the philosophical questions around the existence of a supreme being still stands causing a great disagreement between theists and philosophers as the notion of coherent supreme perfect being is difficult to demonstrate. Moreover, most theists believe that philosophy is not meant for all, which it actually is, but ideas of atheists out of misconception whereas it encapsulates the study of all forms of beliefs due to the fact that it limitlessly encourage the exploration of the understanding of everything using the human power of intellect through logical and rational argument.

Theists are restricted to their beliefs and as such of the opinion that the ontological argument should not be raised at all because it is definitely a threat to all they understand about life.

Science and Philosophy

The journey of how science was extracted from the ancient philosophy has already been narrated. Despite the intimate relationship between them, there seems to be a very major difference that separates them. This difference gave science the right to exist independently as a discipline and makes scientists to explicitly claim that they are different from philosophers. In fact, some have argued that due to this major difference, science has been able to advance and succeed over the years whilst philosophy has been unable to make any significant progress.

The philosophical methodology, that is how to do philosophy and or philosophize, involves a structure of inquiring questions and solving problems associated with everything about life through reasoning and this is taken care of by what is called systematic philosophy. Ancient philosophers like Aristotle, Plato and Spinoza are also regarded as systematic philosophers. However, modern philosophers have developed the concept into various forms like existentialism, analytic philosophy, hermeneutics and deconstructionism. But generally, to philosophize, it starts with the state of being curious as according to Aristotle, "it was their wonder, astonishment, that first led men to philosophize and still leads them", this leads to identifying problem or posing questions to be solved and proffer answers or solutions through theories and analysis which have to be backed up with rational arguments that is open to criticism by other philosophers which most times they do not come into terms. In philosophy, there is not a generally accepted yardstick among philosophers in determining the best theory or explanation to answer a question or solve a problem.

On the other hand, modern scientific methodology is claimed to have originated from a polymath of the medieval age, Ibn al-Haytham (Alhazen) who encouraged the use of experimental data in solving problems but some believed that Aristotle's analytical approach to the logical implication in demonstrative discourse made him the inventor of scientific method. Other figures like Johannes Kepler and Galileo Galilei are also recognized in the development of the scientific method. The scientific method involves a process which also begins with inquiries about an observed phenomenon like; why is the sky blue? Or why does everything that is thrown up comes down? This leads to the formulation of hypothesis to explain the phenomenon and through that, scientists make some predictions which must be testable by an experiment which is analyzed to reach a conclusion and solve the problem or answer the question, thus bring about a science discovery or theory. Mind you, these discoveries or theories can be debunked by a contradicting experimental evidence to cause a new discovery or theory. In short, there is a generally accepted yardstick among scientists used in solving problem. Richard P. Feynman said in one of his lectures:

> "It doesn't matter how beautiful your theory is, it doesn't matter how smart you are. If it doesn't agree with experiment, it's wrong. In that simple statement, is the key to science." [6]

In one statement, the major difference between science and philosophy is their approach towards finding answers to questions associated to life. In modern philosophy, scientific discoveries have been a source of knowledge as such that philosophers interprets them in order to understand everything about life whilst some scientists believe that they can have an adequate understanding of everything about life by studying its fundamental nature, independently of philosophy. This is why most people are of opinion that science have sent philosophy into its shadow and science is all that is necessary to understand all that we want and due to the fact that there is no significant consensus among philosophers causes more problems than solving them.

Science was enjoying its independence and seems to successfully progress in understanding everything about life especially in the medieval age until the 19th century when the novel modern science surfaced. Science, in a more specific context here, is physics. What else if not physics? Since, physics is the central science that study the fundamental nature of everything. Lord Kelvin once said, "In science, there is only physics; all the rest is stamp collecting". So, during the time of Isaac Newton and his contemporaries, classical physics' explanations of the fundamental nature and laws that guide everything was straight and appealing to the intellect, hence, easily understandable. It was easy to understand how the Earth's gravitational force pulls a falling apple, how a ball will only move if a force is applied on it, perhaps by kicking it, such that it will only move in the direction of the force, how it is difficult to drag an object on a rough surface than a smooth surface, why a light through a rain drop will give a rainbow and so on.

Then, modern science (physics) present to us another side of everything with the discoveries of things below the dimension of the physical world around us, the atomic world. The discovery of tiny elementary subatomic particles like electrons, photon, muon, quarks and so on push the quest for understanding what goes on in this micro-world leading to the weirdest aspect of modern science called quantum physics. It required the efforts of several great minds like Neil Bohr, Wiener Heisenberg, Albert Einstein, Richard Feynman, Paul Dirac, Max Planck, Marie Curie, Erwin Schrödinger and just to mention a few.

Quantum physics is weird, strange such that Bohr and his colleagues couldn't believe what they have discovered nor like it but yet, it is the most interesting and fascinating understanding we have ever had about life. However, despite the mysterious nature of quantum physics and the fact that it is difficult to understand, Feynman kind of playfully encouraged everyone not to run away from it while presenting his paper on quantum electrodynamics saying:

> "What I am going to tell you about is what we teach our physics students in the third or fourth year of graduate school…It is my task to convince you not to turn away because you don't understand it. You see my physics students don't understand it. That is because I don't understand it. Nobody does" [7]

The word "quantum" means "smallest possible", hence quantum physics being the physics of the world of small things. Basically, quantum tells us that in the world of the small, nothing is certain and this is referred to as the Uncertainty Principle. For instance, a ball is placed on the floor of a room, classical physics can tell us the exact position of that ball on the floor and if it is in motion, we can calculate the speed at which it is moving and where exactly it will be on the floor of the room at a particular time. In fact, this can be known without looking at the ball at that particular time with mathematical calculations which will give us the exact measurement. But in the quantum world or system, as in the floor of the room should be a quantum floor and we place an electron on it, we cannot tell the exact position of the electron at any particular time but can be anywhere until we look at it. If we are not looking at it, it does not exist at a particular position on the floor but the moment we look at it, it will exist at a particular position i.e. switching from its waveform when we were not looking, to its particle form just the moment we look at it. This whole behavior in the quantum world does not go well with Einstein as he wrote to another physicist, Arnold Sommerfeld in August 1926:

> "The Heisenberg-Dirac theories certainly drive me to admiration, but to me, they don't smell of reality" [8]

The skeptical approach of Einstein towards quantum physics also led to a popular dialogue between him and Bohr which goes thus:

> **Bohr:** I think God throw dice

> **Einstein:** I, at any rate, am convinced that He [God] does not throw dice.

> **Bohr:** Einstein, stop telling God what to do

After all, the strange and unfathomable nature draws the attention of philosophers on the understanding what exactly, quantum physics is trying to tell us about the nature of everything by raising philosophical questions like: What actually happens to a quantum particle when we are not looking at it; What is its exact position and why does it change to a particle when we look at it? Quantum physics seems to bring science back to its origin, philosophy, as it is difficult for scientists to answers these questions without philosophizing. Apparently, quantum physics can be reduced to philosophy by forming a bridge between physics and philosophy and perhaps, this might have been what led Feynman to say that, "Philosophy of science is as useful to scientists as ornithology is to birds".

Therefore, physics and philosophy are not just similar because they both starts with "ph-", actually:

> Physics is Life.
> Philosophy is Life.
> Physics entails the physical body of Life and Quantum Physics can be claimed as its fundamental unit where Philosophy constitutes the soul of Life.

Religion and Science

Science, being an offspring of philosophy, is expected to be quite difficult to integrate with religion. Science believes that through its methodology, the fundamental nature of everything can be understood, described and demonstrated by theories and natural laws. Therefore, more critical in its approach towards understanding everything than the religious perspective which holds some beliefs and convictions of 'divine intervention' by a supernatural being as explained by the divine rules and laws in the sacred scriptures that should or must not be questioned because we can never understand or find an appropriate answer. This creates an unhealthy relationship between Science and Religion. In a TED Talk on militant atheism, an atheist and evolution biologist, Richard Dawkins said:

> "Not only is science corrosive to religion; religion is corrosive to science. It teaches people to be satisfied with trivial, supernatural non-explanations and blinds them to the wonderful real explanations that we have within our grasp. It teaches them to accept authority, revelation and faith instead of always insisting on evidence." [9]

Historically, this unhealthy relationship erupted significantly in the 16th century when Galileo was tried and condemned by the Roman Catholic Inquisition. In 1610, Galileo published his observations which he made through his telescope in his small book, Sidereus Nuncius (Starry Messenger) wherein he promoted the heliocentric theory, that propose the sun to be the center of the universe around which other celestial bodies including the earth revolves, presented by Nicolaus Copernicus in his, De revolutionibus orbium coelestium in 1543. The heliocentric theory is in contrast to the geocentric theory which propose that earth, rather, is the center of the universe around which other celestial bodies, the sun included, revolves. The geocentric model was developed by the Greek astronomer and geographer, Ptolemy which was warmly welcomed by the Catholics as it agrees with the literal interpretation of their scripture. Due to the confliction of Galileo's theory to the geocentric theory, he was tried and prosecuted in 1633 seeing him as a vehement suspect of heresy and therefore sentenced him to indefinite imprisonment as he was placed under house arrest until his death in 1642 after all books on heliocentric theory have been banned.

Apparently, theists or believers are the more offensive side as there are diverse opinions among scientists about theists' belief in divine intervention. Since scientific beliefs are strongly based on experimental evidences plus the significant progress of science in describing the fundamental nature of everything by natural laws and theories, some scientists upheld the opinion that everything must have existed or be existing independently of a divine intervention. While some believe that despite being able to understand the natural laws and theory behind everything, it would not have existed with so much uniqueness and perfection without the effect of a being with absolute intelligence. In 1726, Sir Isaac Newton stated in *Scholium Generale*:

> "The most beautiful system of the sun, planets, and comets, could only proceed from the counsel and dominion of an intelligent and powerful Being" [10]

Moreover, Einstein was of the idea that Science and Religion could still put up with each other and work coherently to have the adequate understanding of everything. He stated in "Science, Philosophy and Religion, A symposium" published by the Conference on Science, Philosophy and Religion in Their Relation to the Democratic Way of Life, Inc., New York in 1941:

> "Accordingly, a religious person is devout in the sense that he has no doubt of the significance and loftiness of those superpersonal objects and goals which neither require nor are capable of rational foundation. They exist with the same necessity and matter-of-factness as he himself. In this sense, religion is the age-old endeavor of mankind to become clearly and completely conscious of these values and goals and constantly to strengthen and extend their effect. If one conceives of religion and science according to these definitions then a conflict between them appears impossible. For science can only ascertain what is, but not what should be, and outside of its domain value judgments of all kinds remain necessary. Religion on the other hand, deals only with evaluations of human thought and action: it cannot justifiably speak of facts and relationships between facts. According to this interpretation, the well-known conflicts between religion and science in the past must all be ascribed to a misapprehension of the situation which has been described." [11]

In fact, Albert has popularly been acknowledged to have said in 1954 that "Science without religion is lame. Religion without science is blind". This ideology also follow suit the theology of Ibn al-Haytham which he described by saying:

> "I constantly sought knowledge and truth, and it became my belief that for gaining access to the effulgence and closeness to God, there is no better way than that of searching for truth and knowledge." [12]

However, there is apparently an equal balance on the debate between theists and scientists before the advent of modern science. Our understanding of the fundamental nature of everything as proposed by classical physics until the 19th century does not provide any scientific evidence to abolish the idea of divine intervention by a supernatural being in the behavior of everything. So, it is more of a fair argument between both sides. Though, the theory of evolution of Charles Darwin seems to explain the origin of biological organisms purely based on his observations in 1859 which suggests that human might have evolved from other species over the years. This was strongly rejected by most theologians who believed in creationism i.e. human was created by God

Nevertheless, the first major clue we had about the how everything came into existence surfaced when Edwin Hubble, in 1929, made some observations with his telescope that the universe is expanding, that is the space between two galaxies are increasing rapidly, which lead to the proposition of the "Big Bang Theory". The theory proposed that at a point earlier in time, everything in the universe was clumped together into a very tiny, hot and dense point or space which expands and cools off to bring about the universe and everything around us. This follows a logical reasoning since the universe is stretching or expanding, then there must have been a point in the past when everything was together at a point. It is important to know that the "Big Bang Theory" only explain the state of everything in the universe at a time in the past, about 13.8 billion years ago and does not in exact suggest the origin of the universe, with no astronomical data to tell us what happened prior to the big bang. So, the question on the origin of the universe was still vague to scientists at this point but obviously, they are very close.

Thoughtfully, it was impossible to propose a theory to explain what happened before that very tiny point of space which supposedly banged to give our universe because at that point, everything will seem to be in singularity, putting it simple and figuratively, that is everything is one and uniform in that tiny, dense and hot space before everything begins to form leading to a Big Bang. Therefore, before the Big Bang at all, there is nothing like space-time which is the basic context on which physical theories are formulated. Space and time which was combined together by Albert Einstein as space-time can be understood such that (to make it simple) it means everything, which is matter behaves by moving through a space in a particular period of time. So, it was thought that the big bang birth the fabric of space-time leading to the creation of the universe and could not obviously have existed before the big bang.

Hence, it is impossible to explain what happened prior to the Big Bang by any theory and it has been dubbed by most physicists as the ultimate boundary or edge of the universe.

Thereafter, about forty years ago, scientists made a significant progress on their understanding of the Big Bang and put forward an explanation like the "inflationary theory" by Alan Guth at Cornell University with significant contribution from Alexei

Starobinsky and Andrei Linde of Landau Institute for Theoretical Physics and Lebedev Physical Institute respectively in the late 1970s and early 1980s which basically suggests that the universe existed from a vacuum space of an infinite gravity by a rapid expansion between the period of 10^{-36} seconds to 10^{-33} or 10^{-32} seconds after the point of singularity before leading to a Big Bang. The inflationary epoch (of rapid expansion) is said to involve a fluctuation of the primordial vacuum space. A weird idea like this could not have make any sense, if not for the crazy physical phenomenon that happens in nature explained by quantum physics. In their experiment on quantum electrodynamics, to study how electrons interact with each other, Richard Feynman and his colleagues discovered a very bizarre phenomenon which changed our intuitive idea of what "nothing" is. He explained that a vacuum, which is thought to be a space containing "nothing", actually, contains a lot of stuffs and events. In a vacuum, sub-atomic particles appears and disappear spontaneously within a very little fraction of second which they termed as 'quantum fluctuations' and this has also been demonstrated in the laboratory several times. However, physicists have proposed that the universe might have resulted from similar process i.e. a universe from nothing. In Tufts University, Professor Alexander Vilenkin developed a variation of the inflationary model which posit that the universe is originated from nothing through quantum tunneling. Quantum tunneling is when a particle disappears at one side of a potential energy barrier and appear at the other side of the barrier just like when we throw a ball over a fence to the other side but instead, we throw the ball at the fence, which tunnel through i.e. pass through the fence (the barrier) to appear at the other side.

This idea was also supported by Paul Davies, an English physicist who is of the thought that the inflationary universe theory that suggests that everything emerges out of nothing can only be as a result of a causeless quantum transition. In the book, God and the New physics, he wrote:

> "In this remarkable scenario, the entire universe simply comes out of nowhere, completely in accordance with the laws of physics, and creates along the way all the matter and energy needed to build the universe as we now see it" [13]

He also wrote further and wonders on the philosophical interpretation of this new idea or proposition by the inflationary model:

> For the first time, a unified description of all creation could be within our grasp. No scientific problem is more fundamental or more daunting than the puzzle of how the universe came into being. Could this have happened without any supernatural input? Quantum physics seems to provide a loophole to the age-old assumption that 'you can't get something from nothing'. Physicists are now talking about the 'self-creating universe': a cosmos that erupts into existence spontaneously, much as a sub nuclear particle sometimes pops out of nowhere in certain high energy processes. The question of whether the details of this theory are right and wrong is not so very important. What matters is that it is now possible to conceive of a scientific explanation for all of creation. Has modern physics abolished God altogether…? [14]

Also, in an essay titled "The Uncaused Beginning of the Universe" by William Lane Craig and Quentin Smith, the latter who is a philosopher is of the idea that there is enough evidence to conclude that the universe began to exist without being caused to do so and the fact about matter is that, we came from nothing, by nothing and for nothing is the most reasonable belief and his writings opines that our universe exist without cause or explanation. If the Big Bang cosmology is right as such exist non necessarily, improbably, and causelessly and in his exact use of words in page 217: *it exists for absolutely no reason at all.* [15]

However, in recent years, an advancement was made based on another application of quantum physics in the understanding of cosmology developed by the famous physicist of the present century, Stephen Hawking, the quantum theory of gravity (combination of quantum mechanics and Einstein's theory of gravity), he explained that time is less fundamental than space i.e. the conjoint of space and time framework, as presented by Einstein, did not exist together at a point during the birth of the universe with 'space' being in existence before 'time' and therefore space-time cannot have a singularity or initial boundary. In other words, the universe did not exist from a point of singularity because there was no singularity. Hence, the universe has no beginning nor end since we can have an atemporal universe. This also implies that our universe came into existence through a bang during the inflationary epoch and not that there was a point of singularity prior to the epoch of inflation. Mind you, no one knows exactly what happened long way back before the inflation, at least as far back as before the 10^{-33} seconds of the inflationary epoch time. In his book, *A Brief History of Time, he wrote*:

> The quantum theory of gravity has opened up a new possibility, in which there would be no boundary to space-time and so there would be no need to specify the behavior at the boundary. One could say: 'The boundary condition of the universe is that it has no boundary.' The universe would be completely self-contained and not affected by

anything outside itself. It would neither be created nor destroyed. It would just BE. [16]

As a result of this, he went ahead to make his theological conclusion in the book:

> So long as the universe had a beginning, we could suppose it had a creator. But if the universe is really completely self-contained, having no boundary or edge, it would have neither beginning nor end: it would simply be. What place, then, for a creator? [17]

Pushing Hawking's argument further, since at that tiny, hot and infinitely dense point of the big bang, the conventional laws of physics as we understand it breaks down due to the infinite gravity, then it follows that, if at a point in the universe or anywhere we can imagine, it becomes so dense and the laws of physics breaks down again, a universe will just be or exist. With this, there will be no end to how many or when the universe can create itself and thus will be eternal. An eternal universe also implies that it will be 'self –contained', for the sake of understanding, just like the law of conservation of energy, energy can never be created nor destroyed but in this case, it will state like, the universe can never be created nor destroyed. In The Grand Design, by Stephen Hawking and Leonard Mlodinow, it is written that:

> "Because there is a law such as gravity, the universe can and will create itself from nothing. Spontaneous creation is the reason there is something rather than nothing, why the universe exists, why we exist. It is not necessary to invoke God to light the blue touch paper and set the universe going." [18]

And more interestingly, it means that our universe is most probably not the only universe in existence but also many other several universes which operate with perhaps different set of laws physics, so to say. This gave rise to the idea of multiverse. However, though the inflationary model is not yet well established and Hawking's idea of a self-contained universe is just a proposal, our new understanding of everything about life has undergone a great paradigmatic shift due to the advent of quantum physics. The surge of the idea that the universe came from nothing, is self-contained and eternally shifts the balance away from the theists and thus, a threat to the religious beliefs as a whole.

The Contingence

Undoubtedly, the religious, scientific and philosophical perspectives of life are systems and concepts driven towards understanding everything that is, which is the first and ultimate zeal of every intelligible human. The validity of the propositions of each perspective have been questioned by other perspectives as explained in the previous discourse, thereby leading to the discrepancies among them. Humans' reactions to these discrepancies is the individual uphold of a perspective over the others with obviously, the negligence of the point where the perspective has failed or the blind belief in the apparently, insufficient explanation from the perspective they have upheld. As a result, humans are unable to realize a unique and unified understanding of everything about life. With this, concerning the understanding of everything that it is, including human himself, we can invariably and safely assert that we have failed. In other words, it is very difficult to perfectly place one perspective over another because due to the limitations and flaws of each of them, it is impossible for one to overwrite the others such that every time we try to place one of the perspectives over the others, we fail miserably.

After all, the principal subject of concern of the three perspectives is life, to understand it. Then it follows an immutable logical reasoning to give a rational conclusion that there can only be one quality for only one subject, that is, in this case, an explanation to understand life can only be valid, invalid or remain in probable state as in 'valid or invalid'. In other words, it is illogical to assert that an explanation is valid and invalid at the same time. Therefore, if the three paradigmatic perspectives of life are valid, then they should be able to share the same space but if they are unable, then obviously there is a loss of connection somewhere causing the differences between them. Following this, we can only be able to bring the three perspectives to share the same space by removing the obstacles i.e. settling the differences between them. This is the only way we can claim that we have succeeded in understanding everything as intelligibles.

Perhaps, bringing the three perspectives to a point of contingence could help to attain a utopian, ideal society which works under a unique and unified understanding of life. This can be likened to Al-Farabi's political philosophy which propose that philosophers have an adequate grasp of essence than believers and therefore a philosopher, as a ruler is the best for a society than a believer because, due to the fact that he possess the essential reality rather than images predicted by religion, he will be able to tolerate all forms of religions in that society. Mind you, the main point here is, Al-Farabi's political philosophy is able to point out that in order to attain an ideal, peaceful society, there have to be a unified worldview that will tolerate all discrepancies just like the philosopher will be able to tolerate all religions of different images, through which it will operate.

So, let us remove the obstacle on our way to our ultimate goal and try to bring together the three major perspectives through which we understand life to a point of contingence by solving the differences between them. It is crazy and seems implausible but a trial is definitely not a crime and moreover, it is very important we do it. So why don't we?

Sequential to the earlier discourse in the previous chapter, our major obstacles are: One, the argument of a coherent supreme perfect being posed by philosophy against the existence of a supreme being, and two, the idea of an eternal, causeless and self-contained universe created from nothing posed by quantum physics on the existence of everything, both which are contradictory to the central religious beliefs. Although, philosophy and science employ different methodology, they share more in common than differences and despite the separation after the medieval age, the advent of quantum physics has strongly revealed and resuscitated the bond between them. The interpretation of the concepts of quantum physics tends to be more in form of philosophical questions and as a result, philosophers and physicists have come up with different interpretations to understand the weird mathematics of the quantum world, like the Copenhagen, Ensemble, Many-worlds interpretations, De-Broglie – Bohm theory and so on. However, the Copenhagen interpretation is one of the oldest and arguably the generally accepted interpretation proposed by Neils Bohr and Werner Heisenberg in 1927.

The uncertainty principle fundamentally tells us that the quantum system is full of uncertainty and contingent events. For better understanding, say an electron in a quantum system like an atom, is in rapid motion as such that it is in a waveform and thus, difficult to know the exact position of the electron (before measurement). In this case, we can only work with probability and predict an area in the atom which is most probable or with the highest probability, where the electron can be found. In other words, every area of the atom has a chance although not equal and so, the electron is said to be in superposition or in lay man term, we say the electron is everywhere in the atom at the same time. Then, after measurement through observation, we know the exact position of the electron in the atom which will, of course, be one of the possible areas in the atom among all. So we ask, what happens to other possible areas and why is the electron not found in other possible areas rather than the one observed? The Copenhagen interpretation says that the wave function of the electron collapses into one possible quantum state after observation and the collapse is caused by the observer. Using the analogy above, this means, the wave form of the electron before measurement collapses into a particle form which is then found in one of the possible areas after observation, in the atom while the other possible areas cancel out. Other interpretations claim that, other possible areas do not cancel out but instead, they are reduced into the one observed or measured one i.e. the probability of having the electron in each of the other possible areas in the atom sum up together to the actual point of area where it is observed. Or according to Hugh Everett's Many-World interpretation in 1957, the probability of having the electron in other possible areas are manifested in another universe i.e. the electron will be in those other possible locations (in the atom) in other universes different from ours.

Nevertheless, the physical theories of quantum physics still stands and is consistent with reality, irrespective of the different interpretations. In other words, it is well grounded that matter is contingent at the quantum level and we can only with certainty, be able to explain the behavior of matter by measurement or observation which results into the transition from a quantum system to a classical system. Therefore the philosophical interpretation can be said to be open to all as long as it follows a logical reasoning and rational argument.

Invariably, quantum physics has bridged the gap between philosophy and science created over the years by connecting them and proved that the combination of the two is necessary to have a better understanding of everything about life.

In spite of this, we move on to address the other major obstacles on our way to the point of contingence, by filling the gaps between two perspectives with the third i.e. using the science of quantum physics to bring philosophy and religious beliefs into terms and settling the difference between the science of quantum physics and religious beliefs by philosophizing with rational arguments. This approach follows a premise that, 'we can only give what we have' which means we can only settle the differences between our paradigmatic perspectives of life with what we already understood about life. It is irrational to think that we can solve a problem without any knowledge, we can only solve a problem with what we already know.

The Eternal, Causeless, Self-Contained Universe

The physical world around us, universe, has been proposed to have come into existence with a bang from a point in an inflationary state of rapid expansion of space, explicitly from 'nothing' due to quantum transitions or fluctuations, as presented by Vilenkin and Guth. It is also said to be 'self-contained' and 'eternal', a proposition by Stephen Hawking with his explanation of the quantum gravity theory which cancels out the notion of an initial singularity. Does this mean the universe is not caused at all? Is the idea of a universe originated from a supernatural being a myth? Hawking categorically claimed in his book, *A Brief History of Time*, that with the advent of the Big Bang quantum cosmology theory, the understanding of how physical world around us came into being or the beginning of the universe has entered the 'realm of science'.

However, the medieval philosopher, Ibn Sina (Avicenna) made a clear distinction between the way of understanding of natural philosophy (science) and metaphysical philosophy. As quoted in A. Hyman and J. Walsh (eds.), *Philosophy in the Middle Ages*, in his *al-Shifa': al-ilahiyyat*, Avicenna wrote:

> "...the metaphysicians do not intend by the agent, the principle of the movement only, as do the natural philosophers, but also the principle of existence and that which bestows existence, such as the creator of the world"
> [19]

This exclusively counters the notion of Hawking as the Big Bang quantum cosmology only present us the principle that brought about the existence of everything but does not in any way tell us what brought about the principle itself. These are not the same thing but two different things that explicitly shows the limit of science. Avicenna also gave a better understanding of what is meant by 'created from' and 'to create' in the same book, as translated in Georges Anawati, *La Métaphysique du Shifa'*, he wrote:

> "This is what it means that a thing is created, that is, receiving its existence from another... As a result everything, in relation to the first cause, is created... Therefore, every single thing, except the Primal One, exists after not having existed with respect to itself." [20]

In the same vein, Avicenna opines that the understanding of essence i.e. what it means for something to exist, reveals what it is, is different from whether it exists and based on this ontological distinction between existence and essence, he asserts that all beings other than God, who expresses this distinction, require a cause to exist.

The cause or from what the state of inflationary epoch, which expands rapidly from a vacuum space of 'nothing', from which the universe exist through a subsequent bang at that tiny big bang point, receive its existence, is not and can never be explained by physicists. Clearly, based on the data and information we have from our universe, we can only know or understand all that happened from the final 10^{-33} seconds of inflation and all that came after. For what happened before then or how long the inflation lasted, we do not know. And in fact, according to a physicist, Ethan Siegel, it is impossible to know exactly what happened before that final time of inflation. Though, Vilenkin and Hawking's theories suggest that the inflation started from a small vacuum space. But the question still stands; what caused that small vacuum space which inflated? If there is no cause for it, does that mean the vacuum space is the First cause and Primal One, thus in that sense, God?

Picking it from Aristotle's perspective, his theory of theology, in understanding what God is, he indirectly termed the supreme being(s) as the 'Unmoved Mover(s)' who is(are) characterized with absolute 'actuality' and devoid of 'potentiality' who moves the 'Movers' that characterized with some 'potentiality'. In this context, Aristotle's notion of movement means 'motion'. According to him, the Unmoved Movers actualizes the potentiality in the movers which he exemplify with the stars, planets, celestial bodies and all material things, that is, the Unmoved Movers determines their motion whilst they don't themselves, undergo any form of motion due to their actuality nature. In this case, the vacuum space can definitely not be the first cause because it is characterized with 'potentiality' which is actualized, setting it into motion to inflate and thereby, bring about the universe. Definitely, the vacuum space has to be caused, even though we don't know exactly the cause.

Thomas Aquinas, a Christian philosopher whose works was influenced by Aristotle, also described the existence of God in his works, "Summa contra Gentiles" and "Summa theologiae" by posing the five renowned arguments called *quinque viae* (Five Ways): the argument of motion, argument of causation, argument of existence of necessary and the unnecessary, argument of gradation and argument of ordered tendencies of nature.

- **The argument of motion**: Just like the idea of Aristotle, there are some things that definitely undergo motion or move which cannot be caused by themselves except another (a mover). The chain of caused mover cannot be infinite according to Aquinas and therefore, there has to be a 'First Mover' who is not caused to move by any other whom is what everyone understands as God.

- **The argument of causation:** Following the argument of motion, a thing is cause by a thing which is also caused by another and the chain goes on. This cannot lead to an infinite regress and so there has to be a First Cause which understood to be God.

- **The argument of existence of necessary and the unnecessary:** Aquinas argues that some things exist in our experience that are unnecessary but not everything can be unnecessary. Therefore, something have to exist necessarily, with the necessity only from itself and as such is itself the cause for other things to exist.

- **The argument of gradation:** Things that exist are in gradation, like good, cold, hot, colder, hotter, better and so on. But there has to be a thing of the highest grade, truest and noblest thing fully existing which we call God.

- **The argument of ordered tendencies of nature**: By nature, things that obeys the natural laws undergo actions that are ordered towards an end i.e. have a beginning and end. Then, there has to something that is does not have a beginning and end to direct these things to achieved the order of nature.

It should be noted, Aquinas' arguments doesn't prove affirmatively that God exists but only describe the 'existence of God'.

Using his argument as a premise, the vacuum space could not be consider as the First Mover because it must have been set into motion to expand by a mover, therefore caused to be, will be unnecessary without being caused, is not of the highest gradation since it is subjected to "change" under certain laws of nature. In spite of these, the Big Bang quantum cosmology theory of a universe from nothing is not sufficient to abolish the existence of God. Aquinas also gave an account on the nature of God to define his divine qualities which he expressed by considering what God is not, *via negative*, as thus: God is simple, perfect, infinite, immutable and one without diversification. Simple in the sense that he does not compose of matter or form, perfect due to his pure actuality, infinite as different from the created beings, immutable as he does not change in his level of essence and one as his essence is its existence i.e. his essence and existence are unified into one.

Moreover, Avicenna argues that the real existence that is acknowledged to a being was already present as an essence or possibility in the divine mind and therefore an additional benefit blessed to the possible being by God in the act of creation and this was develop by Aquinas – as stated in David Burrell's book, *Aquinas and Islamic and Jewish Thinkers* – to give the idea of radical dependency such that the creaturely existence is understood not as something which happens to essence but as a fundamental relation to the Creator as origin. All of these points out that, the understanding of the essence of creation of the universe as explained by the Big Bang quantum cosmology does not give an account on its origin, that cause its existence.

In lieu of all that, since the vacuum space could not have been the First Mover, then what caused the vacuum space into existence? This is a question that we cannot or never be able to answer with apt scientific evidences or data as pointed earlier. Hawking did proposed an answer which is, perhaps the "vacuum space" was just a tiny part another prior existing space or universe which could have also existed from another vacuum and it infinitely goes on like that. So, our universe could just be one out of many universes which also existed from a bang in an inflating space from nothingness. If Hawking is right, a universe is probably being created with a bang somewhere far away at this very moment. It's crazy but that is what his proposition implies; an "eternal" universe.

However, the argument of an eternal universe had surfaced earlier in the medieval ages of philosophy when Avicenna argues that the universe is infinite and eternal causing a rift between the philosophers and theologians present then. This led to the rise of Al-Ghazzali, a philosopher and theologian, who argues against the idea in his book, *The Incoherence of the Philosophers*. Theologians like Al-Ghazzali believe that the universe is created and the notion of an eternal universe as viewed by the philosopher is different to their orthodoxical position of belief. This requires the intervention of Averroes (Ibn Rush'd) who reconciled the difference between them in his book, *Decisive Treatise*, by pointing out that the misunderstanding is simply as a result of misconception in the definition, wherein he stated that:

> "As for the question whether the world is eternal or has been generated, the disagreement between the Ash'arite dialectical theologians and the ancient sages almost comes back, in my view, to a disagreement about naming, especially with respect to some of the Ancients. That is because they agree that there are three sorts of existing things: two extremes and one intermediate between the extremes. And they agree about naming the two extremes but disagree about the intermediate". [21]

Ibn Rush'd approached the dispute by understanding the points of agreement and disagreement between philosophers and theologians. As indicated in the citation above, philosophers and theologians agree that there are three basic classes of beings. On the two extremes of the classes of beings, philosophers and theologians are obviously in terms but disagrees on the intermediate class which Averroes believe that, with the right definition, it can be reconciled. He explained in *Decisive Treatise* thus:

> "One extreme is an existent thing that exists from something other than itself and by something—I mean, by an agent cause and from matter. And time precedes—I mean, its existence. This is the case of bodies whose coming into being is apprehended by sense perception, plants, and so forth. The Ancients and the Ash'arites both agree in naming this sort of existing things 'generate." [22]

> "The extreme opposed to this is an existent thing that has not come into existence from something or by something and that time does not precede. About this, too, both factions agree in naming it 'eternal.' This existent thing is apprehended by demonstration; it is God…" [23]

> "The sort of being between these two extremes is an existent thing that has not come into existence from something and that time does not precede, but that does come into existence by something—I mean, by an agent. This is the world as a whole." [24]

The problem with the intermediate class arise due to the similar qualities it shared with both extremes. In common with God, it does not come into existence from something i.e. from nothing and not preceded by time while in common to the other extreme class, its existence is caused by something. However, Averroes believe that the problem can only be solved by balancing the similarity of the world to the opposite extremes. He stated in the same book:

> "So it is evident that this latter existent thing has been taken as resembling the existing thing that truly comes into being and the eternally existing thing. Those overwhelmed by its resemblance to the eternal rather than to what is generated name it 'eternal,' and those overwhelmed by its resemblance to what is generated name it 'generated.' But, in truth, it is not truly generated, nor is it truly eternal. For what is truly generated is necessarily corruptible, and what is truly eternal has no cause" [25]

Therefore, Ibn Rush'd redefined the intermediate class, although more closely to Avicenna's understanding of essence and described the world to be created from nothing and eternally caused. In Greek Essence and Islamic Tolerance: Al-Farabi, Al-Ghazzali, Ibn Rush'd by Michael Sweeney, Averroes' description is explained this way:

> The intermediate class, however, that Ibn Rush'd describes is actually much closer to Ibn Sina's understanding of essence than it is to his own more Aristotelian approach. For Ibn Sina, (a) the existence of essence is caused, and (b) it is not produced from matter, although it is produced from essence as possibility, and it is (c) eternally caused to be by God's essential activity. At the very least, Ibn Rush'd's own view is that matter is eternal and uncreated; if the world comes to be, it comes to be from matter. Avicennian essence is a potentiality that is not matter, and thus it better reconciles the philosophical view that the world cannot come to be from nothing with the religious view that God does not create the world from a preexisting matter". [26]

For a better understanding, what Ibn Rush'd's description is trying to tell is that the eternal essence of the universe, which to Avicenna, is a potentiality or possibility, is caused to exist by God due to his own essence which is equal to his eternal existence. In spite of these, the Big Bang cosmology that explains the existence of the universe from nothing by quantum fluctuations which is self-contained, conserved and eternal as proposed by the quantum gravity theory does not necessarily affirm that the universe is causeless and thus not make the essence of a supernatural being unnecessary. By this, we have been able to settle the difference between Science and religious perspectives. Sorry to Prof. Hawking and his fellow fans of the 'eternal, self-contained universe' from this side.

Coherent Supreme Perfect Being

The absolute perfection attributed to God by Descartes i.e. the idea of coherent supreme perfect being such that he is omniscient, omnipresent and omnipotent at the same time was criticized by Leibniz and Kant who felt that it is difficult to demonstrate. This is a major obstacle we have in reconciling philosophy and the religious notion of the existence of a Supreme Being. However, a lot has happened over the years and our understanding of everything has undergone a great paradigm shift, most importantly with the advent of modern science. As explained earlier in the previous chapter, the bizarre, absurd and mysterious nature of quantum mechanics has unarguably revealed the secrets behind the nature of everything and as such raised some philosophical question about life, giving room for philosophical interpretations. Perhaps our new mysterious understanding of quantum physics can help to make religion and philosophy see eye to eye.

So we pose the question, can quantum physics give a demonstration of a coherent perfect supreme being? Well, let's give it a try, shall we?

The weirdest and most counter-intuitive of all the phenomenon of physical nature described by quantum physics is "quantum entanglement". Quantum entanglement is a theoretical framework in which a pair or group of particles are entangled i.e. connected together, having an effect on each other even when they are separated by a very large distance. For instance, an electron with an angular spin of zero (0) gives two photons (particles of light) and they are entangled with each other. If one photon is placed on the north pole of the Earth and the other photon is placed on the south pole of the Earth, then you measure the spin of the one on the north pole, say it is clockwise ($\frac{1}{2}$), then it will instantaneously effect the other one on the south pole to give its own measurement, anticlockwise (-$\frac{1}{2}$), even without necessarily looking or observing it. Note that the measurement of the photons adds up to the spin of the electron ($\frac{1}{2}$-$\frac{1}{2}$ = 0), i.e. entanglement also follows the law of conservation such that the quantum state of each entangled particle (photons) in the system must add up to the quantum state of the whole system (electron, in this case). Moreover, the spin of the entangled particles is not the only parameter we can measure, it also goes for other parameters too, like the position, polarization, energy, momentum and velocity of a quantum particle. So using energy instead of spin in the instance above with the law of conservation, that means, the energy state of each entangled quantum particle must add up to the energy state of the whole system (both together).

Neils Bohr, based on the fact that the laws of quantum physics are based on nothing but probability and uncertainty, explained that before an entangled particle is observed, there is a probability for all the possible quantum states the particle can be in the system i.e. in the analogy above, there is probability that the photon (North or South before observation) could spin clockwise just as there is a probability that it could spin anticlockwise. In other words, the quantum states of the entangled particle is uncertain before observation and thus, it is in superposition, meaning, all the possible quantum state are superposed or more explanatorily, using the case of the two photons, the photons are spinning clockwise and anti-clockwise at the same before observation. So after observation, the wave function of the particle collapses which instantaneously effects its entangled partner, collapsing its wave function too in order to give a measurement. This idea was inconceivable for Albert Einstein and he clearly disliked the whole thing presented by the elegant mathematics of quantum physics. He, outrightly claimed that the theory of quantum physics is, though not wrong, but incomplete because he felt that it is impossible for a particle to instantaneously effect another, which seems to occur at a speed faster than the light speed. In other words, going against the speed limit as stated by classical physics. In 1935, Einstein and two others, Boris Poldosky and Nathan Rosen presented a paper to show that the quantum theory is incomplete by formulating the thought experiment known as Einstein-Poldosky-Rosen Paradox (EPR Paradox) which in simple terms, argued that the quantum state of each entangled particle is predetermined i.e. certain, even before they are observed, so the entangled particles are not in superposition as claimed by Bohr. They wrote in the paper:

> "We are thus forced to conclude that the quantum-mechanical description of physical reality given by wave functions is not complete" [27]

For a better understanding, Einstein's argument can be explained with the glove thought experiment. Say, we have a pair of gloves, right and left hand and we put each glove in a box such that the two boxes are entangled. Then the boxes are separated and one is placed somewhere on the moon while the other is placed on an island on Earth. If we open the box on the island and we observe a right hand glove, then obviously, the box on the moon will definitely contain a left hand glove. In other words, the type of glove in each box is already predetermined and not in any way probabilistic. To Einstein, he felt this is the best explanation to describe entanglement without defying the classical speed limit. He later described entanglement as '*spukhafte Fernwirkung*' i.e. spooky action at a distance. Erwin Schrödinger was also not comfortable with the whole idea of entanglement and wrote in one of his paper, Discussion of probability relations between separated systems in 1935:

> "I would not call [entanglement] one but rather the characteristic trait of quantum mechanics, the one that enforces its entire departure from classical lines of thought." [28]

Also in the same year, 13th of July, he wrote to Einstein saying:

> "I know of course how the hocus pocus works mathematically. But I do not like such theory" [29]

In 1955, Einstein died without any conviction about his thought on entanglement and with the firm belief that quantum theory is rather incomplete, so his debate with Bohr still lives on. In 1967, at Columbian university, John Clauser who was pursuing his PhD in Astrophysics decided to take up the challenge. He clearly disliked the whole quantum mechanics idea because he could not even understand it, so he felt, Einstein must have been right.

Through his course of research, he stumbled on a paper by an Irish physicist, John Stewart Bell, who described a way to settle the debate between Einstein and Bohr. He posited that if Bohr's quantum mechanical understanding of entanglement is wrong, then the whole quantum theory is not just incomplete but totally wrong. Therefore, he stated in his paper on the Einstein-Poldosky-Rosen paradox, he theorized that the debate can be settled if a machine can be built which can create and compare many pairs of entangled particles. He turned the whole question of understanding entanglement into an experimental question. The first experimental breakthrough was from Carl Kocher in 1967 who built an apparatus which emitted two photons from a calcium atom that were shown to be entangled. Clauser with the contribution of Stuart Freedman in 1972 built a better apparatus of Kocher's apparatus equipped with better polarizers which was able to create and compare thousand pairs of entangled particles. As the results were coming in from their observations, Clauser was shocked and unhappy because the experimental data proved Einstein wrong. In other words, Bohr's quantum mechanical explanation was apt and definitely, an entangled particle is in superposition before observation and instantaneously effect its partner the moment it is observed. Stephen Hawking has been reported to have said: "Einstein was confused, not quantum mechanics". Alas! This is a physical phenomenon that occurs over a wide range of distance faster than the speed of light and thus defying the law of classical physics. Mind you, the instantaneous effect of an entangled particle on its partner after observation does not involve any transfer of energy, signal or information between them but rather, it just be. So what does this mean?

In spite of this, perhaps this new counter-intuitive picture of the nature of reality we now understand, presented by quantum physics can help to describe the concept of a coherent Supreme perfect being. Quantum physics make us to realize that the fundamental nature of reality i.e. everything about life is guided by uncertainty and probability, which means, there is a probability, no matter how small, that everything (life) can present us some paranormal, miraculous or supernatural events. In other words, quantum physics is trying to make us understand that there is at least, a slight chance that a thing, matter, can behave in a manner that is completely different from the way we usually understand it to behave. Using the analogy of the entanglement explained above, an entangled particle is in a superposed state of all the possible quantum states it can be (like the photon spinning clockwise and anti-clockwise simultaneously), before observation and there is a probability for every quantum state (clockwise or anti-clockwise) that will be observed. So, the way we understand a particular thing to usually behaves, is just one of the numerous quantum state that can be observed and the moment it behaves unusual to our normal experience, thus kind of miraculously, we have only observed a different quantum state among all the possible states. Understanding this and connecting it with the religious ideas, a supreme being must be omnipotent i.e. possess unlimited power to cause even the most unusual event or miracles. Then, if quantum physics is telling us, in principle, that there are possibilities for the occurrence of unusual and miraculous events, then why is it hard for us to believe that these events are caused by an omnipotent being? This was expressed by Bryan Enderle on Science and God in this manner:

"If modern science tells that there are examples of unusual activities with finite probabilities, then theologically, we are not as surprised by miracles which may merely be low probability events" [30]

This can also be understood in another context presented in the works of the earlier philosophers, especially Avicenna and Al-Ghazzali, on the notion of essence. For Avicenna, the essence of a thing is a possibility or potentiality that is actualized i.e. caused by God, through his own essence which is the only essence that does not need a cause whilst other essences of the thing that are not caused remain as possibilities. In his book, The *Metaphysics of The Healing* translated by Michael Marmura, he stated that:

"We will now return to what we were [discussing] and say: There are specific properties that belong individually to each of the Necessary Existent and the possible existent. We thus say: The things that enter existence bear a [possible] twofold division in the mind. Among them there will be that which, when considered in itself, its existence would be not necessary. It is [moreover] clear that its existence would also not be impossible, since it would not enter existence. This thing is within the bound of possibility. There will also be among them that which, when considered in itself, its existence is necessary."

"If its quiddity is sufficient for either of the two states of affairs [existence or nonexistence] to obtain, then that thing would be in itself of a necessary quiddity, when [the thing] has been supposed not to be necessary [in itself]. And this is contradictory. If [on the other hand] the existence of its quiddity is not sufficient [for specifying the possible with existence]—[the latter] being, rather, something whose existence is added to it—then its existence would be necessary due to some other thing. [This,] then, would be its cause." [31]

In the citation above, what Avicenna is trying to point out is that, the essence of a thing is a possibility that can exist or not but will only exist if it is necessary for it to exist in order for the thing to be what it is. Avicenna posit further that the essence of a thing that exist is something that the human intellect is able to conceive which means any other essence of a thing that are inconceivable by human intellect cannot exist thus cannot be caused. Therefore, the Supreme Being is only limited to the necessary essence and cannot cause other essence inconceivable to the human intellect. For instance, if we look at cotton and fire, the essence of cotton is to be burned when in contact with fire and in the reverse case, the essence of the fire is to burn the cotton when in contact with it. So in Avicenna explanation of essence, which is in accordance with that of Plato and Aristotle, the phenomenon of a cotton, not burning to ash when in contact with fire, cannot exist because it is inconceivable to human intellect likewise the cotton burning to ash without introducing to fire.

However, Al-Ghazzali criticized Avicenna's understanding of essence and points out that he had joined the 'cause' and the 'effect' together which instead, should be seen as two different things. That is, Avicenna has attached "the cause of burning" to the "fire" whereas the effect of the essence of the fire being caused is the "burning" just like the turning of the cotton into ash when in contact with the fire is also an effect of the essence of the cotton being caused. Since, the essence of the fire (to burn) and cotton (to turn to ash) are possibilities that have to be caused, then God, being omnipotent, out of his own essence to cause, can independently cause the essences of the fire and cotton. As quoted in Michael Sweeney's *Greek Essence and Islamic*

Tolerance: Al-Farabi, Al-Ghazzali, Ibn Rush'd, Al-Ghazzali express the problem of essence thus:

> First, essence as actually existing is caused to be, but essence as possible—essence as essence—is uncaused. Essence as possible is the "matter" from which God creates the world through intermediaries; in other words, essence is an intelligibility and a metaphysical possibility that is given to God as well as to the human intellect. Second, God is limited by the causal necessities inherent in essence. For Ibn Sina, when the essences of fire and cotton meet, burning is a necessary effect. God's causality ultimately makes the essences of fire and cotton to be rather than not to be, but he cannot make fire and cotton to act other than according to their essences, and thus God is neither a creator ex nihilo nor omnipotent. [32]

Joining all of these together with the understanding of quantum physics, using the analogy of the cotton and fire and in simple terms, the essence of a thing are like the quantum states i.e. either the cotton turns to ash, without any contact with fire, with contact with fire or not turning to ash even with contact with fire are three different essences (quantum states) of the cotton which are three different possibilities that can be actualized (observed) as caused by an omnipotent Supreme Being. So if quantum physics tells us that any of the possible quantum states (essence) can be observed then it is not hard to believe that unusual events are caused by an omnipotent being. Another example is "fire" and "water", water will quench a fire due to combustion, like the fire from a burning coal or wood while in the case of an electrical fire, it will fuel the fire instead. The two essences of the fire to "quench due to water" and "continue burning in the presence of water" are two different probabilities just as the two essences of water to "quench" or "fuel" a fire are also two different probabilities and these, four probabilities, can be caused independently and not necessarily dependently on another even if they are of low chance.

Al-Ghazzali even made an interesting progress in his understanding of essence by proposing that other essences that are not actualized which does not exist and remain as possibilities could exist in another possible worlds even though he did not like this idea. He claimed that the essence of a thing that exist, only exists because it necessary for this particular world. In other words, he had indirectly followed Everett's interpretation of quantum mechanics of many-world theory. In a book by Frank Griffel, *Al-Ghazali's Philosophical Theology*, he wrote:

> "For Avicenna, there can be no world alternative to the one that exists. . . [A]l-Ghazali wishes to express that God could have chosen to create an alternative world in which the causal connections are different from those of this world. Al-Ghazali upholds the contingency of the world against the necessitarianism of Avicenna. For al-Ghazali, this world is the contingent effect of God's free will and His deliberate choice between alternative worlds" [33]

All in all, quantum superposition of states give room for probabilities that everything can actualize some unusual events thus demonstrating that it could be caused by an omnipotent supreme being. With that done, we go on to discover if quantum physics can also help us with the idea of omnipresence and omniscient.

The notion of the omnipresence characteristic nature of the Supreme Being is a way in which believers or theists answer the question of where God is. Theists believe that God is everywhere and with his omnipotence nature of being, he is able to interact and cause everything in life simultaneously. Looking closely, quantum entanglement could help us to understand this sort of perfect nature of the Supreme Being. Quantum entanglement tells us that, as explained earlier, two or more entangled particles can interact with each other over a very wide a range of distance, even without anything connecting them. So, if that is, then why is it hard for us to believe that, an omnipresent Supreme Being, can interact with everything simultaneously as seem he is everywhere? Bryan Enderle also expressed this saying:

> "If modern science tells us that particles can interact over variable distances, then we are not as surprised that God can interact with humans over variable distances" [34]

However, that is not all. Entanglement also involves another weird concept that, perhaps can help us to demonstrate the omniscient nature of God. By omniscient, theists mean, God experiences the past, present and future at the same time and thus, All-Knowing. Another way to understand this is to say God is outside of time because to Him, nothing like past, present and future as he experience the three together and simultaneously. So, entanglement made us to realize that the observation of one of the entangled particles instantaneously effect the others such that it occurs at a speed faster than light. Now, if we borrow Einstein's most renowned classical theory of relativity which tells us that the ultimate speed limit for any object is the speed of light which is constant at 3×10^8 m/s (this was what made that genius dude disagree with the Bohr's idea of quantum entanglement after all). In other words, light is the fastest thing in the universe. Pushing it further, the theory also made us to realize that the faster we move, the faster the rate at which time lags or slow down until we reach the maximum speed (speed of light). At this maximum speed, time exist no more i.e. when light travels at its speed, it is outside of time or does not experience time. For photons (the particles of light), there is no time and so they experience past, present and future at the same time.

In 1971, an experiment was made by a physicist, Joseph C. Hafele and an astronomer, Richard E. Keating to test the theory of relativity called, Hafele-Keating experiment. They took a couple of cesium beam atomic clocks and synchronized them, then place one on the ground at the United States Naval Observatory and took the other aboard commercial airliners which was flew around the world in different directions, eastward and westward. Thereafter, they brought the clocks together and compare them and realized that they are no more synchronized such that their differences were consistent with the predictions of the theory of relativity.

In spite of this, it is evident that time slows down the faster we go and at the speed of light, we are said to be out of time. So, putting everything together, the concept of quantum entanglement tells us that instantaneous events occurs outside of time since it occurs at a speed faster than the speed of light. If quantum physics is telling us this, then it should not be hard that for us to believe that God, being omniscient, can be outside of time. Also, according to Bryan Enderle:

> "If modern science tells us that light can be outside of time, then we are not as surprised that God may be outside of time" [35]

Therefore, quantum physics has been able to demonstrate to us the possibility of how a being can possess the perfect characteristics of omniscience, omnipotence and omniscience. But there is still one more important thing.

The main argument of Descartes and Leibniz is the demonstration of the coherency of these characteristics in one being for absolute perfection. Well, this is not farfetched, the quantum entanglement phenomenon is enough to demonstrate the coherence as it involves the three major concepts needed, observation based on uncertainty and probability, interaction over a wide range of distance without any connection and instantaneous effect that occurs outside of time in just one quantum system. Invariably, a quantum system of entanglement demonstrate that three characteristics of omnipotence, omniscience and omnipresence can be unified in a single system and thus affirms possible coherency of the three in a supreme being. Example of naturally entangled systems can be observed in atoms with multi-electrons which are entangled in their shells, the process of photosynthesis have also been reported to involve the concept of entanglement in harnessing the energy from sunlight and also been observed in living organisms like bacteria in their relationship with quantized light. With this, we have been able to settle the difference between religious beliefs and philosophy and thereby bringing them together to share the same space.

The Unified Paradigm

The point of contingence of the idea of, the existence of God central to the religious beliefs, our understanding of the crazy nature of reality, of everything about life explained by the absurd science of quantum physics and our intuitive curiosity nature of understanding everything about life based on the power of our intellect, is that point that tolerates all these three, that forms our major perspectives of life, without any contradiction. Although, the point does not necessarily mean that the three, perfectly and appropriately in terms with each other but rather, can still share the same space harmoniously without any major conflicting contradictions.

It is very important we understand that this text has not explicitly, with certainty, proven that a Supreme Being, God, exists or everything and the universe as a whole was certainly caused. It has only been able to create and maintain a balance among these assertions and arguments. In other words, the fact that it has been able to counter the notion of a causeless and self –contained universe as a result of our understanding of the Big Bang quantum cosmology through rational argument does not affirmatively prove that the universe is definitely caused to exist as it has not as well been able to proffer scientific evidence with empirical data that the universe was caused. All it has done is, philosophize logically and appropriately, to interpret and have an accurate understanding of the Big Bang quantum cosmology.

In the same vein, describing a coherent supreme perfect being with the understanding of quantum physics does not imply that it has been able to proffer a direct evidence of the existence of God as it is just a proof by analogy, neither does it imply that God is a quantum particle. All it has done is to maintain a balance between philosophy and religious beliefs such that there is no contradiction between them and they will be able to tolerate each other. Furthermore, the description of coherency in nature of God does not in any way rescue the problem of "free will", as this is still a major conundrum in the world of Philosophy and Science.

Nevertheless, a direct proof of evidence for the existence of God as a being independent of the universe, is quite impossible because we don't have direct knowledge of God's nature i.e. there is no any account from anyone with direct experience of what God is in exact, is like or made up of, as all we know or understand is everything in the physical world around us and nothing beyond. Therefore, we can only indirectly proffer evidence for his existence as this text has done. In the same vein, it is too early to conclude that the universe existence is only driven by the supposedly universal principle or law of quantum gravity, independent of any other being, as it still requires a sounder foundation of empirical evidences.

Nonetheless, by clearing the contradiction between philosophy, modern science of quantum physics and religious beliefs, it have been able to unify them into one paradigm perspective, that tolerates them harmoniously, through which we can understand life. The unified paradigm perspective of life is such that it tells us that our understanding of the nature of everything about life does not in any way disprove, affirmatively, the existence of a Supreme Being who caused everything, even though there is not any direct approach, in principle, to understand the nature of the supreme being Himself nor does it affirm that everything in the physical world around us just be without a cause. So the question of if God exists and how everything came into existence i.e. either caused or uncaused to exist is still, logically and in principle, an open question. In another sentence, even despite the critical argument presented by philosophers and the success of scientists in understanding the principle of everything around us, this has not been able to thrash the idea of the existence of God while on the other hand, theologians are only able to give, indirectly, an account of the existence and essence of God as a supreme being and unable, in principle, to proffer a direct account of his existence and essence. With this new unified paradigm shift, we've unarguably made a great success as humans and intellectuals in our understanding of everything as it can, perhaps help us to attain an ideal utopian society.

In spite of these, the claim of the existence of God – a primal being or existence which influences everything – is the most probable, even though we are unable to give a direct account, with apt logical explanation of his nature, but just as Anslem of Canterbury posited, *since we can conceive Him in mind, thus He exists.*

---

[1] Philosophy of Plato and Aristotle, part I: The Attainment of Happiness. Al-Farabi.
[2] Ibid
[3] Aristotelian theology, Wikipedia.
[4] Al-asfar al-arba'a, Vol. 6, pp. 14-16. Mulla Sadra.
[5] Descartes, Rene. Meditations on First Philosophy V: On the Essence of Material Objects and More on God's Existence.
[6] Feynman's Lecture on "The Character of Physical Law", Cornell University, 1964.
[7] Feynman, Richard QED: The Strange Theory of Light and Matter.
[8] The Collected Papers of Albert Einstein. Vol 15: The Berlin Years: Writings & Correspondence, June 1925-May 1927 (English Translation Supplement). Edited by Diana Karmos Buchwald, Jozsef Illy, A.J. Kox, Dennis Lehmkuhl, Ze'ev Rosenkranz & Jennifer Nollar James, pp. 196

[9] Richard Dawkins on militant atheism, TED talk, 2007.

[10] Isaac Newton; General Scholium, Philosophiae Naturalis Principia Mathematica, 3rd Edition, 1726.

[11] "Albert Einstein: Religion and Science". Sacred-texts.com. Retrieved 2013-06-16, Relationship between religion and science, wikipedia.com.

[12] Plott 2000, Pt. II, p.465

[13] God and the New Physics (New York: Simon and Schuster, 1983), p. 215

[14] ibid. p. viii.

[15] William Lane Craig and Quentin Smith, op. cit., p. 217.

[16] Hawking, A Brief History of Time, op. cit., p. 136

[17] ibid., p. 141

[18] Micheal Holden (2010-09-02). "God did not create the universe, says Hawking" Reuters.com. Retrieved 2010-10-17, The Grand Design, Wikipedia.

[19] al-Shifa': al-Ilahiyyat, VI. 1, quoted in A. Hyman and J. Walsh (eds.), Philosophy in the Middle Ages, second edition (Hackett, 1983), p. 248.

[20] al-Shifa': al-Ilahiyyat, VIII.3, translated in Georges Anawati, La Métaphysique du Shifa' (Paris, 1978), Vol. II, pp. 83-84

[21] Ibn Rush'd, Decisive Treatise (Provo, Utah; Brigham Young University), p. 14

[22] Ibid. 14

[23] Ibid. 15

[24] Ibid. 15

[25] Ibid. 15

[26] Michael Sweeney; Greek Essence and Islamic Tolerance: Al-Farabi, Al-Ghazzali, Ibn Rush'd, p. 55

[27] Kumar, Manjit (2011). Quantum: Einstein, Bohr, and the Great Debate about the Nature of Reality. (Reprint ed.). W.W. Norton & Company. pp. 305-306. ISBN 978-0393339888.

[28] Schrodinger Erwin (1935). "Discussion of probability relations between separated systems". Mathematical Proceedings of the Cambridge Philosophical Society. 31 (4): 555-563

[29] Elise Crull (06-07-2019). "If You Thought Quantum Mechanics Was Weird, You Need to Check Out Entangled Time". Sciencealert.com

[30] Bryan Enderle, TED talk, Science and God, 2013

[31] Avicenna, The Metaphysics of The Healing trans. by Michael Marmura. (Provo, Utah; Brigham Young University, 2005) Book 1, Chapter 6, pp. 29-30

[32] Michael Sweeney; Greek Essence and Islamic Tolerance: Al-Farabi, Al-Ghazzali, Ibn Rush'd, p.46

[33] Frank Griffel; Al-Ghazali's Philosophical Theology (Oxford, Oxford University

Press, 2009), p. 173

[34] Bryan Enderle, TED talk, "Science and God", 2013.

[35] Ibid.

www.ingramcontent.com/pod-product-compliance
Lightning Source LLC
Chambersburg PA
CBHW070507220526
45467CB00002B/600